What Has

by Ellen Catala

Table of Contents

Consultant: Dwight Herold, Ed.D., Past President,
Iowa Council for the Social Studies

Everything Changes

As time goes by, you and everything
in your world changes. You are not
the same as you were a few years ago.
You are growing up! The **inventions**
around you are changing, too.

What did people do before these things were invented? How have these inventions changed over time? Let's look at a few of the important inventions in our world.

Candles to Electric Lights

Before there were electric lights, people used candles or gas lamps to see at night. These did not give off much light.

Then the **lightbulb** was invented. Soon it was much easier to work, shop, read, and play after dark. The world became a brighter place.

Letters to Telephones

For years, people could only "talk" long distance by writing letters. Then, in 1876, Alexander Graham Bell invented the telephone.

After a while, telephone lines connected friends and family all over the country. Today, we have **cell phones**—no wires needed!

Horses to Cars

How did people get around in the old days? By horse, of course! It took days to travel from one city to another, and the roads were narrow and muddy.

Then the automobile, or car, was invented. Travel became easier, and the streets were made wider and smoother. Over time, cars got even better and faster!

Fires to Ovens

Long ago, people cooked food over an open fire. It was hard, hot work. Then, the gas stove was invented. It made cooking much easier.

As people went from using gas to using electricity, the electric stove came along. Today, many people also use **microwave ovens**. They cook things very fast!

Paintings to Photographs

Long ago, it was not possible to take a picture of someone. You had to draw or paint it. Then the camera was invented.

At first, people had to sit still for several minutes to have their **photograph** taken. Then cameras got better, and taking pictures became a snap!

Feet to Bicycles

It's probably hard for you to imagine a world without bicycles, but that's how it was for a long time. The early bicycles looked a little different from the bicycles we ride today.

Early bicycles did not have padded seats or brakes. These changes made biking more comfortable and safer. Today, bicycles are even better!

Glossary

cell phone   a wireless telephone that uses radio waves to send and receive calls

invention   a new device or machine that is made based on an idea someone has

lightbulb   a glass bulb with a filament inside that glows and gives off light when electricity flows through it

microwave oven   small electric oven that uses radio waves to cook food quickly

photograph   a picture taken by a camera using light-sensitive film